Antimicrobial Therapy and Clinical Bacteriology

Diagnosis and Treatment of Infectious Diseases

Bhupan Thapa

Copyright © 2024 Bhupen Thapa
All rights reserved.
No part of this book may be reproduced or transmitted in any form or by any means, electronic or mechanical, including photocopying, recording, or by any information storage and retrieval system, without permission in writing from the copyright holder.

Table of Content

Antimicrobial Therapy and Clinical Bacteriology 1
Diagnosis and Treatment of Infectious Diseases 1
   Table of Content 2
   Chapter 1 1
   Introduction to Clinical Bacteriology 1
      The Role of Bacteriology in Medicine 1
      History of Clinical Bacteriology 2
      Importance of Bacterial Identification in Diagnosis and Treatment 3
   Chapter 2 5
   Bacterial Structure and Function 5
      Cell Wall Composition 5
      Membrane Structure 6
      Bacterial Reproduction and Growth 7
   Chapter 3 9
   Laboratory Techniques in Clinical Bacteriology 9
      Specimen Collection and Transport 9
      Microscopy and Staining Techniques 10
      Culture and Sensitivity Testing 11
   Chapter 4 13
   Common Bacterial Pathogens 13
      Gram-Positive Bacteria 13
      Gram-Negative Bacteria 14
      Atypical Bacteria 15
   Chapter 5 17
   Clinical Manifestations of Bacterial Infections 17
      Localized Infections 17
      Systemic Infections 18
      Opportunistic Infections 19
   Chapter 6 21
   Diagnosis of Bacterial Infections 21
      Clinical Presentation 21
      Laboratory Tests 22
      Molecular Diagnostics 23
   Chapter 7 25
   Treatment of Bacterial Infections 25
      Antibiotics and Antibiotic Resistance 25
      Antimicrobial Therapy 26
      Prophylactic Measures 27
   Chapter 8 29

Prevention and Control of Bacterial Infections 29
    Vaccines 29
    Infection Control Measures 30
    Public Health Strategies 31
Chapter 9 33
Emerging Trends in Clinical Bacteriology 33
    Antibiotic Stewardship 33
    Genomic Epidemiology 34
    Novel Therapeutic Approaches 35
Chapter 10 37
Case Studies in Clinical Bacteriology 37
    Case Study 1: Pneumonia 37
    Case Study 2: Urinary Tract Infection 38
    Case Study 3: Sepsis 39
Chapter 11 41
Future Directions in Clinical Bacteriology 41
    Technological Advancements 41
    Global Health Implications 42
    Research Priorities 43
Conclusion: The Future of Clinical Bacteriology 44
References 45

A Primer for Healthcare Professionals

Chapter 1

Introduction to Clinical Bacteriology

The Role of Bacteriology in Medicine

Bacteriology plays a crucial role in the field of medicine by focusing on the study of bacteria and their effects on human health. Understanding the role of bacteriology in medicine is essential for diagnosing and treating infectious diseases effectively. Bacteria are microscopic organisms that can cause a wide range of illnesses, from minor infections to life-threatening diseases. By studying the characteristics and behavior of bacteria, healthcare professionals can better understand how to prevent, diagnose, and treat bacterial infections.

One of the key roles of bacteriology in medicine is in the diagnosis of infectious diseases. Bacteriologists use a variety of techniques to identify and characterize bacteria present in clinical samples, such as blood, urine, or tissue samples. By isolating and identifying the specific bacteria causing an infection, healthcare providers can determine the most effective treatment options, such as antibiotics or antiviral medications. Additionally, bacteriology plays a critical role in monitoring the spread of infectious diseases and identifying outbreaks before they become widespread.

In addition to diagnosis, bacteriology also plays a crucial role in the development of vaccines and other preventative measures against bacterial infections. By studying the structure and behavior of bacteria, scientists can develop vaccines that stimulate the immune system to recognize and attack specific bacteria, preventing infection or reducing its severity. Bacteriology also contributes to the development of strategies for infection control and prevention in healthcare settings, helping to reduce the spread of antibiotic-resistant bacteria and other dangerous pathogens.

Furthermore, bacteriology is important in understanding the mechanisms of antibiotic resistance and developing new antibiotics to combat resistant bacteria. Antibiotic resistance is a growing global health threat, as bacteria evolve to resist the effects of commonly used antibiotics. Bacteriologists study the genetic mechanisms of antibiotic resistance in bacteria and work to develop new drugs that can effectively treat infections caused by resistant strains. By staying ahead of the curve in antibiotic development, bacteriology plays a vital role in ensuring that healthcare providers have the tools they need to combat infectious diseases effectively.

Overall, the role of bacteriology in medicine is essential for understanding, diagnosing, and treating infectious diseases. By studying the behavior and characteristics of bacteria, healthcare professionals can develop effective strategies for preventing and treating bacterial infections. Bacteriology also plays a critical role in the development of vaccines, infection control measures, and new antibiotics to combat antibiotic resistance. As our understanding of bacteriology continues to evolve, so too will our ability to combat infectious diseases and protect public health.

History of Clinical Bacteriology

The history of clinical bacteriology is a fascinating journey that has revolutionized the field of medicine and our understanding of infectious diseases. The roots of clinical bacteriology can be traced back to the 19th century, when pioneering scientists such as Louis Pasteur and Robert Koch made groundbreaking discoveries that laid the foundation for modern microbiology.

Louis Pasteur is often credited as the father of microbiology for his work on the germ theory of disease, which proposed that microorganisms are the cause of many infectious diseases. His experiments with fermentation and sterilization techniques paved the way for the development of vaccines and antiseptics, which have saved countless lives over the years.

Robert Koch, on the other hand, is best known for his work on identifying the specific bacteria that cause diseases such as tuberculosis, cholera, and anthrax. Koch's postulates, a set of criteria used to establish a causal relationship between a microorganism and a disease, are still used today in the field of clinical bacteriology to identify and study pathogenic bacteria.

In the late 19th and early 20th centuries, the field of clinical bacteriology continued to advance rapidly with the discovery of antibiotics such as penicillin by Alexander Fleming and the development of techniques for culturing and identifying bacteria in the laboratory. These advancements have revolutionized the diagnosis and treatment of infectious diseases, allowing for more targeted and effective therapies.

Today, clinical bacteriology plays a crucial role in the field of medicine, helping healthcare professionals diagnose and treat a wide range of infections caused by bacteria. From simple urinary tract infections to life-threatening sepsis, the knowledge and techniques developed through the history of clinical bacteriology continue to save lives and improve the health outcomes of patients around the world.

Importance of Bacterial Identification in Diagnosis and Treatment

The identification of bacteria plays a crucial role in the diagnosis and treatment of infectious diseases. Understanding the specific type of bacteria causing an infection is essential for determining the most effective course of treatment. Different bacteria respond to different antibiotics, so accurate identification is necessary to ensure that the appropriate medication is prescribed. Inaccurate identification can lead to ineffective treatment, prolonged illness, and even the development of antibiotic resistance.

Bacterial identification is also important for preventing the spread of infectious diseases. Knowing the exact strain of bacteria involved allows healthcare professionals to implement appropriate infection control measures to prevent the spread of the disease to others. This is particularly important in hospital settings, where patients are more vulnerable to infections and where resistant strains of bacteria can easily spread.

In addition to guiding treatment and preventing the spread of infection, bacterial identification is crucial for monitoring the emergence of antibiotic resistance. By identifying the specific bacteria causing an infection, healthcare professionals can track the prevalence of resistant strains and adjust treatment protocols accordingly. This information is essential for developing new antibiotics and strategies to combat antibiotic resistance.

Advances in technology have made bacterial identification faster and more accurate than ever before. Molecular techniques, such as polymerase chain reaction (PCR) and next-generation sequencing, allow for rapid and precise identification of bacteria, even in complex clinical samples. These tools have revolutionized the field of clinical bacteriology, enabling healthcare professionals to diagnose infections more quickly and accurately than ever before.

Overall, the importance of bacterial identification in diagnosis and treatment cannot be overstated. Accurate identification is essential for guiding treatment decisions, preventing the spread of infection, monitoring antibiotic resistance, and ultimately improving patient outcomes. Healthcare professionals must continue to prioritize bacterial identification in order to effectively combat infectious diseases and protect public health.

Chapter 2

Bacterial Structure and Function

Cell Wall Composition

Cell wall composition is a crucial aspect of bacterial biology that plays a significant role in the pathogenicity and antibiotic resistance of various bacterial species. The cell wall is a rigid structure that surrounds the bacterial cell membrane and provides structural support and protection. The composition of the cell wall varies among different bacterial species and is a key determinant of their susceptibility to antibiotics.

The cell wall of bacteria is mainly composed of peptidoglycan, a polymer made up of repeating units of N-acetylglucosamine and N-acetylmuramic acid cross-linked by short peptide chains. Peptidoglycan provides strength and rigidity to the cell wall, allowing bacteria to maintain their shape and resist osmotic pressure. The thickness and composition of the peptidoglycan layer vary among different bacterial species, influencing their susceptibility to antibiotics that target cell wall synthesis.

In addition to peptidoglycan, the cell wall of some bacteria may also contain other components such as lipopolysaccharides, teichoic acids, and proteins. These additional components play various roles in bacterial physiology and pathogenicity. For example, lipopolysaccharides are important virulence factors in Gram-negative bacteria and play a role in immune evasion and host-pathogen interactions.

The composition of the cell wall can also influence the effectiveness of antibiotics against bacterial infections. Antibiotics such as penicillin and cephalosporins target the synthesis of peptidoglycan, making them more effective against bacteria with thick peptidoglycan layers. On the other hand, bacteria with altered cell wall composition or the ability to modify their cell wall structure may exhibit resistance to these antibiotics, posing challenges for treatment.

Understanding the composition of the bacterial cell wall is essential for the development of effective antibiotics and treatment strategies for bacterial infections. By targeting specific components of the cell wall, researchers can design novel therapeutics that are more effective against resistant bacterial strains. Additionally, knowledge of cell wall composition can help clinicians predict the susceptibility of bacterial infections to different antibiotics, guiding treatment decisions and improving patient outcomes.

Membrane Structure

Membrane structure is a crucial aspect of bacterial cells that plays a vital role in their survival and function. The bacterial cell membrane, also known as the plasma membrane, is a thin, flexible barrier that surrounds the cell and separates its internal environment from the external environment. This membrane is composed of a phospholipid bilayer, which consists of two layers of phospholipid molecules arranged in such a way that their hydrophobic tails are facing each other, while their hydrophilic heads are facing the aqueous environments both inside and outside the cell.

In addition to phospholipids, the bacterial cell membrane also contains proteins that are embedded within the lipid bilayer or attached to its surface. These membrane proteins play a variety of roles, including transport of molecules across the membrane, signal transduction, and cell adhesion. Some membrane proteins also serve as receptors for external signals, allowing the cell to respond to changes in its environment.

The bacterial cell membrane is selectively permeable, meaning that it allows only certain molecules to pass through while blocking others. This selective permeability is crucial for the survival of the cell, as it allows the cell to control the movement of substances in and out of the cell. Small molecules such as oxygen and carbon dioxide can freely diffuse across the membrane, while larger molecules and charged particles require the assistance of transport proteins to cross the membrane.

The structure of the bacterial cell membrane is also important for its ability to maintain the cell's shape and integrity. The phospholipid bilayer provides a stable barrier that protects the cell from its surroundings, while the proteins embedded within the membrane help to maintain the cell's shape and allow it to interact with other cells and its environment. Disruption of the cell membrane can have serious consequences for the cell, leading to loss of cell integrity and function.

Overall, understanding the structure and function of the bacterial cell membrane is essential for studying bacterial physiology and for developing new strategies for treating bacterial infections. By targeting the cell membrane, researchers and clinicians can identify new drug targets and develop novel antimicrobial agents that can effectively disrupt bacterial cell membrane function and eradicate bacterial infections.

Bacterial Reproduction and Growth

Bacterial reproduction and growth are fundamental processes that play a crucial role in the development and spread of infectious diseases. Understanding these processes is essential for diagnosing and treating bacterial infections effectively. In this subchapter, we will explore the mechanisms of bacterial reproduction and growth, as well as the factors that influence these processes.

Bacterial reproduction occurs through a process called binary fission, where a single bacterial cell divides into two identical daughter cells. This process allows bacteria to rapidly increase in number under favorable conditions. The rate of bacterial growth is influenced by various factors, including nutrient availability, temperature, pH, and oxygen levels. Bacteria require specific nutrients, such as carbohydrates, proteins, and vitamins, to fuel their metabolic processes and support their growth.

The growth of bacteria can be divided into four distinct phases: lag phase, exponential phase, stationary phase, and death phase. During the lag phase, bacteria adapt to their environment and prepare for rapid growth. The exponential phase is characterized by a period of intense bacterial growth, where the population doubles in number with each generation. In the stationary phase, the growth rate of bacteria slows down as nutrients become depleted and waste products accumulate. Finally, in the death phase, the number of bacterial cells decreases as the environment becomes inhospitable for growth.

Factors such as antibiotic use, immune system response, and environmental conditions can affect bacterial reproduction and growth. Antibiotics work by targeting specific pathways involved in bacterial growth, such as cell wall synthesis or protein production. However, bacteria can develop resistance to antibiotics through mechanisms such as mutation or horizontal gene transfer. Understanding the mechanisms of bacterial reproduction and growth is essential for developing effective diagnostic tools and treatment strategies for infectious diseases caused by bacteria.

Chapter 3

Laboratory Techniques in Clinical Bacteriology

Specimen Collection and Transport

Specimen collection and transport play a crucial role in the accurate diagnosis of infectious diseases in clinical bacteriology. Proper collection of specimens ensures that the laboratory receives a high-quality sample for testing, which is essential for obtaining reliable results. In this subchapter, we will discuss the importance of specimen collection and transport in the context of diagnosing bacterial infections.

When collecting specimens for bacterial culture, it is important to use appropriate collection devices and techniques to prevent contamination. For example, sterile containers should be used to collect specimens such as urine, blood, and cerebrospinal fluid, while swabs are commonly used for collecting specimens from the throat, nose, and wounds. Additionally, healthcare providers should follow strict aseptic techniques to minimize the risk of introducing external bacteria into the sample.

Proper labeling of specimens is also essential to ensure accurate identification and tracking throughout the testing process. Each specimen should be labeled with the patient's name, date and time of collection, type of specimen, and any relevant clinical information. This information is crucial for the laboratory to accurately process and report results back to the healthcare provider.

After collection, specimens should be transported to the laboratory in a timely manner to prevent degradation of the sample and ensure accurate test results. Some specimens, such as urine and stool samples, may require refrigeration or other specific handling instructions to maintain their integrity during transport. It is important for healthcare providers to follow the recommended guidelines for specimen transport to ensure the reliability of test results.

In conclusion, proper specimen collection and transport are essential steps in the diagnosis and treatment of infectious diseases in clinical bacteriology. By following appropriate collection techniques, labeling specimens correctly, and transporting them to the laboratory in a timely manner, healthcare providers can help ensure the accuracy and reliability of bacterial culture results. These steps are crucial for guiding the appropriate treatment of bacterial infections and ultimately improving patient outcomes.

Microscopy and Staining Techniques

Microscopy and staining techniques are essential tools in the field of bacteriology, allowing for the visualization and identification of bacterial cells. One of the most commonly used microscopy techniques in bacteriology is light microscopy, which uses visible light to magnify and illuminate bacterial cells. By using a light microscope, bacteriologists can observe the size, shape, and arrangement of bacteria, providing valuable information for diagnosis and treatment.

In addition to light microscopy, fluorescence microscopy is another powerful tool in bacteriology. This technique involves the use of fluorescent dyes that bind to specific bacterial structures, allowing for the visualization of these structures under ultraviolet light. Fluorescence microscopy is particularly useful for identifying specific bacterial species or structures, as the fluorescent dyes can be targeted to bind to unique characteristics of the bacteria.

Staining techniques are also commonly used in bacteriology to highlight specific bacterial structures and aid in identification. One of the most widely used staining techniques is the Gram stain, which classifies bacteria into two groups based on their cell wall structure. Gram-positive bacteria retain the purple stain, while Gram-negative bacteria appear red after staining. This simple staining technique is a quick and effective way to differentiate between different types of bacteria.

Another important staining technique in bacteriology is acid-fast staining, which is used to identify bacteria that have a waxy cell wall, such as Mycobacterium tuberculosis. Acid-fast staining involves the use of a special dye called carbol fuchsin, which binds to the waxy cell wall of the bacteria and resists decolorization by acid-alcohol. This technique is crucial for the diagnosis of tuberculosis and other diseases caused by acid-fast bacteria.

Overall, microscopy and staining techniques play a crucial role in the field of bacteriology, allowing for the visualization and identification of bacterial cells. By utilizing these techniques, bacteriologists can accurately diagnose and treat infectious diseases caused by bacteria, ultimately improving patient outcomes and public health.

Culture and Sensitivity Testing

Culture and sensitivity testing is a crucial step in the diagnosis and treatment of infectious diseases caused by bacteria. This testing involves isolating the bacteria from a clinical specimen, such as blood, urine, or sputum, and growing it in a laboratory setting. Once the bacteria have been cultured, they are then subjected to various antibiotics to determine which ones are effective in killing the bacteria. This information is vital in guiding the selection of the most appropriate antibiotic therapy for the patient.

The culture part of the testing process involves placing the clinical specimen onto a growth medium that provides the necessary nutrients for the bacteria to thrive. The medium is then placed in an incubator set at the optimal temperature and humidity for bacterial growth. Bacteria will begin to multiply and form colonies on the medium, which can then be identified and tested for sensitivity to antibiotics.

Once the bacteria have been identified, sensitivity testing is performed to determine which antibiotics are effective in inhibiting or killing the bacteria. This is typically done using a method called disc diffusion, where paper discs containing different antibiotics are placed on the agar plate with the bacterial colonies. The plate is then incubated, and the size of the zone of inhibition around each disc is measured to determine the sensitivity of the bacteria to each antibiotic.

The results of culture and sensitivity testing are crucial in guiding the selection of antibiotic therapy for the patient. By identifying which antibiotics are most effective against the bacteria causing the infection, healthcare providers can tailor treatment to maximize efficacy and minimize the development of antibiotic resistance. This personalized approach to antibiotic therapy is essential in ensuring the best possible outcome for the patient.

In summary, culture and sensitivity testing play a vital role in the diagnosis and treatment of infectious diseases caused by bacteria. By isolating and identifying the bacteria causing the infection, as well as determining their sensitivity to antibiotics, healthcare providers can make informed decisions about the most appropriate antibiotic therapy for the patient. This personalized approach to treatment is crucial in ensuring the best possible outcome and minimizing the development of antibiotic resistance.

Chapter 4

Common Bacterial Pathogens

Gram-Positive Bacteria

Gram-positive bacteria are a diverse group of microorganisms that play a significant role in infectious diseases. These bacteria have a thick cell wall composed of peptidoglycan, which gives them their characteristic purple stain when subjected to Gram staining. This group includes well-known pathogens such as Staphylococcus aureus, Streptococcus pyogenes, and Clostridium difficile, among others.

One of the key features of Gram-positive bacteria is their ability to produce toxins that contribute to their pathogenicity. For example, Staphylococcus aureus produces a variety of toxins that can cause skin infections, pneumonia, and toxic shock syndrome. Streptococcus pyogenes produces toxins that contribute to conditions such as strep throat and necrotizing fasciitis. Understanding the toxins produced by these bacteria is crucial for diagnosing and treating infections caused by them.

Gram-positive bacteria are also known for their resistance to antibiotics. Some strains of Staphylococcus aureus, for example, have developed resistance to multiple antibiotics, making them difficult to treat. This highlights the importance of appropriate antibiotic stewardship practices to prevent the further development of antibiotic-resistant strains of Gram-positive bacteria.

In clinical bacteriology, identifying the specific species of Gram-positive bacteria causing an infection is essential for selecting the appropriate treatment. This often involves performing culture and sensitivity testing to determine which antibiotics are effective against the bacteria. In some cases, molecular techniques such as polymerase chain reaction (PCR) may be used to quickly identify the bacteria and guide treatment decisions.

Overall, a thorough understanding of Gram-positive bacteria is essential for diagnosing and treating infectious diseases caused by these microorganisms. By recognizing their unique characteristics, including their ability to produce toxins and develop antibiotic resistance, healthcare professionals can effectively manage infections and prevent the spread of resistant strains.

Gram-Negative Bacteria

Gram-negative bacteria are a diverse group of microorganisms that are characterized by the structure of their cell walls. These bacteria have a thin peptidoglycan layer surrounded by an outer membrane, which contains lipopolysaccharides. This outer membrane is responsible for many of the unique properties of gram-negative bacteria, including their resistance to certain antibiotics and their ability to cause severe infections in humans.

One of the most well-known gram-negative bacteria is Escherichia coli, which is a normal resident of the human gut but can also cause infections such as urinary tract infections and food poisoning. Other gram-negative bacteria include Pseudomonas aeruginosa, which is a common cause of hospital-acquired infections, and Neisseria gonorrhoeae, which causes the sexually transmitted infection gonorrhea.

Gram-negative bacteria are often more resistant to antibiotics than gram-positive bacteria due to the presence of the outer membrane, which acts as a barrier to the entry of certain drugs. This can make treating infections caused by gram-negative bacteria more challenging and may require the use of multiple antibiotics or alternative treatment strategies.

In addition to their antibiotic resistance, gram-negative bacteria are also known for their ability to produce toxins that can cause severe illness in humans. For example, certain strains of E. coli produce a toxin that can cause kidney failure, while Pseudomonas aeruginosa produces toxins that can damage lung tissue and lead to pneumonia.

Overall, understanding the unique characteristics of gram-negative bacteria is essential for diagnosing and treating infections caused by these microorganisms. By studying their structure, antibiotic resistance mechanisms, and toxin production, healthcare providers can develop more effective strategies for managing infections and preventing their spread in clinical settings.

Atypical Bacteria

Atypical bacteria, also known as "unculturable bacteria," are a group of microorganisms that do not grow well on standard laboratory culture media. These bacteria are difficult to detect using traditional methods and often require specialized techniques for isolation and identification. Despite their elusive nature, atypical bacteria play a significant role in infectious diseases and can cause a wide range of illnesses in humans.

One of the most well-known atypical bacteria is Mycoplasma pneumoniae, a common cause of community-acquired pneumonia. Unlike typical bacteria, Mycoplasma lack a cell wall, making them resistant to many antibiotics that target cell wall synthesis. This unique characteristic poses a challenge for clinicians when selecting appropriate treatment options for Mycoplasma infections.

Another important atypical bacterium is Chlamydophila pneumoniae, which is known to cause respiratory infections such as bronchitis and pneumonia. Chlamydophila pneumoniae is an obligate intracellular pathogen, meaning it can only replicate inside host cells. This intracellular lifestyle allows the bacterium to evade the immune system and makes it difficult to target with conventional antibiotics.

Legionella pneumophila is another atypical bacterium that can cause severe respiratory infections, including Legionnaires' disease. This bacterium is commonly found in natural water sources, such as rivers and lakes, and can be transmitted to humans through inhalation of contaminated water droplets. Legionella infections can be challenging to diagnose due to the bacterium's fastidious nature and the need for specialized testing methods.

In conclusion, atypical bacteria present a unique challenge for clinicians due to their unusual characteristics and the specialized techniques required for their detection and treatment. Understanding the biology and pathogenicity of atypical bacteria is essential for accurate diagnosis and effective management of infections caused by these microorganisms. By staying informed about the latest advancements in laboratory techniques and treatment options, healthcare providers can improve patient outcomes and reduce the morbidity associated with atypical bacterial infections.

Chapter 5

Clinical Manifestations of Bacterial Infections

Localized Infections

Localized infections are a common occurrence in the field of bacteriology, often presenting with specific symptoms and diagnostic challenges. These infections are typically confined to a specific area of the body, such as the skin, respiratory tract, or urinary tract. Understanding the mechanisms of localized infections is crucial for accurate diagnosis and effective treatment.

Skin infections are among the most common localized infections encountered in clinical practice. These infections can be caused by a variety of bacteria, including Staphylococcus aureus and Streptococcus pyogenes. Common skin infections include cellulitis, abscesses, and impetigo. Diagnosis of skin infections often involves clinical examination, along with laboratory tests such as wound cultures or Gram staining to identify the causative organism.

Respiratory tract infections, such as pneumonia and sinusitis, are another common type of localized infection. These infections are typically caused by respiratory pathogens like Streptococcus pneumoniae, Haemophilus influenzae, and Legionella pneumophila. Diagnosis of respiratory tract infections may involve imaging studies like chest X-rays, along with respiratory secretions analysis and culture to identify the specific bacteria causing the infection.

Urinary tract infections (UTIs) are also frequently encountered localized infections, particularly in women. The most common causative agents of UTIs are Escherichia coli, Klebsiella pneumoniae, and Proteus mirabilis. Diagnosis of UTIs involves urine analysis and culture to identify the presence of bacteria and determine the appropriate antibiotic therapy. Treatment of localized infections often involves antibiotics, along with supportive care to manage symptoms and prevent complications.

In conclusion, localized infections are a significant aspect of clinical bacteriology, requiring a thorough understanding of the specific pathogens involved and the best diagnostic and treatment approaches. By recognizing the signs and symptoms of localized infections and utilizing appropriate laboratory tests, healthcare providers can effectively diagnose and manage these infections to ensure optimal patient outcomes.

Systemic Infections

Systemic infections are a serious concern in the field of bacteriology, as they can have widespread effects on the body and pose a significant risk to the health of the patient. These infections occur when bacteria enter the bloodstream and spread throughout the body, causing symptoms such as fever, chills, fatigue, and body aches. Systemic infections can be caused by a variety of bacteria, including Staphylococcus aureus, Streptococcus pneumoniae, and Escherichia coli.

Diagnosing systemic infections can be challenging, as the symptoms can be nonspecific and overlap with other conditions. However, there are several key diagnostic tests that can help identify the presence of bacteria in the bloodstream. Blood cultures are a common test used to detect the presence of bacteria in the blood, while other tests such as complete blood counts and inflammatory markers can provide additional information about the severity of the infection.

Treatment for systemic infections typically involves the use of antibiotics to target the specific bacteria causing the infection. The choice of antibiotic will depend on the type of bacteria identified through diagnostic testing, as well as the severity of the infection and the patient's overall health. In some cases, intravenous antibiotics may be necessary to ensure effective treatment and prevent the spread of the infection to other organs.

In addition to antibiotics, supportive care may also be necessary to help manage the symptoms of systemic infections and support the body's immune response. This can include measures such as fluid resuscitation, pain management, and monitoring for complications such as sepsis or organ failure. Close monitoring of the patient's condition is essential to ensure timely intervention and prevent further complications.

Overall, systemic infections present a significant challenge in the field of bacteriology, requiring prompt diagnosis and appropriate treatment to prevent serious complications. By understanding the causes, symptoms, and treatment options for systemic infections, healthcare providers can better manage these complex and potentially life-threatening conditions.

Opportunistic Infections

Opportunistic infections are a common concern in clinical bacteriology, particularly in patients with compromised immune systems. These infections occur when normally harmless microorganisms take advantage of a weakened immune system to cause disease. This can happen in individuals with conditions such as HIV/AIDS, cancer, diabetes, or those undergoing immunosuppressive therapy. Understanding opportunistic infections is crucial for accurate diagnosis and treatment in these vulnerable populations.

One of the most well-known opportunistic infections is Pneumocystis pneumonia (PCP), caused by the fungus Pneumocystis jirovecii. PCP is a common cause of severe respiratory illness in immunocompromised individuals, particularly those with HIV/AIDS. Diagnosis of PCP typically involves identifying the organism in respiratory specimens using specialized staining techniques. Treatment usually involves the use of antibiotics such as trimethoprim-sulfamethoxazole.

Another opportunistic infection of concern is candidiasis, caused by the yeast Candida species. Candidiasis can manifest as oral thrush, vaginal yeast infections, or invasive bloodstream infections in immunocompromised individuals. Diagnosis of candidiasis often involves culturing the organism from clinical specimens and identifying it based on its characteristic morphology. Treatment may include antifungal medications such as fluconazole or amphotericin B.

Mycobacterium avium complex (MAC) is another opportunistic infection that affects individuals with weakened immune systems, particularly those with advanced HIV/AIDS. MAC infections typically involve the lungs, lymph nodes, or bloodstream, and can be challenging to diagnose due to the slow growth of the organism in culture. Treatment usually involves a combination of antibiotics such as clarithromycin and ethambutol.

Overall, opportunistic infections present unique challenges in clinical bacteriology due to their association with compromised immune systems. Understanding the common pathogens involved, as well as the appropriate diagnostic and treatment strategies, is essential for providing optimal care to patients at risk for these infections. By staying informed on the latest developments in the field of clinical bacteriology, healthcare professionals can better manage opportunistic infections and improve patient outcomes.

Chapter 6

Diagnosis of Bacterial Infections

Clinical Presentation

Clinical presentation refers to the signs and symptoms that a patient exhibits when infected with a bacterium. These can vary widely depending on the type of bacteria involved, the site of infection, and the overall health of the individual. It is essential for healthcare providers to be able to recognize these clinical presentations in order to accurately diagnose and treat bacterial infections.

One common clinical presentation of bacterial infections is fever. This is the body's natural response to infection, as the elevated temperature helps to kill off the invading bacteria. Fever can be accompanied by other symptoms such as chills, sweats, and fatigue, which can help healthcare providers narrow down the potential causes of the infection.

Another common clinical presentation of bacterial infections is inflammation. This can manifest as redness, swelling, pain, and warmth at the site of infection. Inflammation is the body's way of trying to contain and eliminate the bacteria, but it can also cause discomfort and functional impairment for the patient. Recognizing these signs of inflammation can help healthcare providers pinpoint the source of the infection and determine the appropriate treatment.

In some cases, bacterial infections can cause more severe clinical presentations, such as sepsis. Sepsis is a life-threatening condition in which the body's response to infection causes widespread inflammation and organ dysfunction. Patients with sepsis may exhibit symptoms such as rapid heartbeat, rapid breathing, confusion, and low blood pressure. Early recognition and treatment of sepsis are crucial to improving outcomes for these patients.

Overall, understanding the clinical presentation of bacterial infections is essential for healthcare providers in diagnosing and treating these conditions. By recognizing the signs and symptoms of infection, healthcare providers can initiate appropriate testing, prescribe the right antibiotics, and monitor the patient's progress. This knowledge is crucial for providing optimal care for patients with bacterial infections and preventing the spread of these dangerous pathogens.

Laboratory Tests

Laboratory tests play a crucial role in the diagnosis and treatment of infectious diseases. In the field of clinical bacteriology, these tests are essential for identifying the causative agents of infections and determining the most effective treatment options. By analyzing samples taken from patients, laboratory professionals can provide valuable information to healthcare providers, leading to more accurate diagnoses and better patient outcomes.

One of the most common laboratory tests used in clinical bacteriology is the culture and sensitivity test. This test involves growing bacteria from a patient sample in a controlled environment, such as a petri dish, and then exposing the bacteria to various antibiotics to determine which ones are most effective at killing the bacteria. This information is critical for selecting the right antibiotic therapy for the patient, as it helps to avoid the development of antibiotic resistance and ensures the best possible outcome.

Another important laboratory test in clinical bacteriology is the polymerase chain reaction (PCR) test. This test allows for the rapid detection of bacterial DNA in patient samples, providing a quick and accurate diagnosis of bacterial infections. PCR testing is often used in cases where traditional culture methods are not practical or when rapid results are needed for immediate treatment decisions.

In addition to culture and sensitivity tests and PCR testing, laboratory professionals may also perform serological tests to detect antibodies produced by the immune system in response to a bacterial infection. These tests can help confirm a diagnosis and monitor the progress of treatment. Other specialized tests, such as nucleic acid amplification tests (NAATs) and immunoassays, may also be used to identify specific bacterial pathogens or detect the presence of bacterial toxins.

Overall, laboratory tests are essential tools in the field of clinical bacteriology, providing valuable information for the diagnosis and treatment of infectious diseases. By utilizing a combination of traditional culture methods, molecular techniques, and serological tests, laboratory professionals can help healthcare providers make informed decisions about patient care, leading to better outcomes for individuals affected by bacterial infections.

Molecular Diagnostics

Molecular diagnostics have revolutionized the field of clinical bacteriology by providing rapid, accurate, and sensitive methods for detecting and identifying bacterial pathogens. These techniques utilize the genetic material of bacteria to diagnose infections, determine antibiotic resistance patterns, and monitor the spread of infectious diseases. By targeting specific genes or sequences, molecular diagnostics can quickly and efficiently identify the presence of bacteria in clinical samples, enabling healthcare providers to make informed treatment decisions.

One of the most commonly used molecular diagnostic methods in clinical bacteriology is polymerase chain reaction (PCR). PCR amplifies specific regions of bacterial DNA, allowing for the detection of even small amounts of bacteria in patient samples. This technique is highly sensitive and can differentiate between closely related bacterial species, making it invaluable for diagnosing infections caused by multiple pathogens or antibiotic-resistant strains. PCR assays are also faster than traditional culture-based methods, providing results in hours rather than days.

Another important molecular diagnostic tool is nucleic acid sequencing, which involves determining the genetic code of bacterial DNA. By sequencing the entire genome or specific regions of bacterial DNA, researchers can identify unique genetic markers that differentiate between different bacterial strains or species. This information is crucial for tracking the spread of infectious diseases, understanding the mechanisms of antibiotic resistance, and developing targeted therapies for bacterial infections.

In addition to PCR and nucleic acid sequencing, other molecular diagnostic techniques used in clinical bacteriology include fluorescent in situ hybridization (FISH), nucleic acid amplification tests (NAATs), and gene expression profiling. These methods offer a range of benefits, including increased sensitivity, specificity, and speed compared to traditional culture-based tests. By leveraging the power of molecular biology, healthcare providers can quickly and accurately diagnose bacterial infections, optimize antibiotic therapy, and improve patient outcomes.

In conclusion, molecular diagnostics have transformed the field of clinical bacteriology by providing advanced tools for detecting, identifying, and monitoring bacterial pathogens. These techniques offer unparalleled sensitivity and specificity, enabling healthcare providers to make timely and informed decisions about patient care. As technology continues to advance, the role of molecular diagnostics in clinical bacteriology will only continue to grow, driving improvements in the diagnosis and treatment of infectious diseases.

Chapter 7

Treatment of Bacterial Infections

Antibiotics and Antibiotic Resistance

Antibiotics have revolutionized the field of medicine by providing effective treatments for bacterial infections. These medications work by killing or inhibiting the growth of bacteria, helping to alleviate symptoms and promote recovery. However, the overuse and misuse of antibiotics have led to the development of antibiotic resistance, a growing concern in the medical community.

Antibiotic resistance occurs when bacteria evolve and become resistant to the effects of antibiotics. This can occur through genetic mutations or the transfer of resistance genes between bacteria. As a result, infections caused by resistant bacteria are more difficult to treat and may require stronger or more expensive antibiotics. In some cases, there may be no effective treatment available, leading to prolonged illness or even death.

To combat antibiotic resistance, it is essential for healthcare providers to prescribe antibiotics judiciously. This includes only prescribing antibiotics when necessary, choosing the appropriate medication, and ensuring that patients complete the full course of treatment. Patients can also help by following their healthcare provider's instructions, avoiding the use of leftover antibiotics, and never sharing medications with others.

In addition to prudent antibiotic use, research into new antibiotics and alternative treatments is crucial for addressing antibiotic resistance. Scientists are working to develop novel antibiotics that target different mechanisms of bacterial growth, making it more difficult for bacteria to develop resistance. Other approaches, such as phage therapy and immunotherapy, offer promising alternatives to traditional antibiotic treatments.

Overall, the issue of antibiotic resistance is complex and multifaceted, requiring collaboration between healthcare providers, researchers, policymakers, and the general public. By working together to promote responsible antibiotic use, support research into new treatments, and raise awareness about the importance of combating antibiotic resistance, we can help preserve the effectiveness of antibiotics for future generations.

Antimicrobial Therapy

Antimicrobial therapy plays a crucial role in the treatment of infectious diseases caused by bacteria. It involves the use of medications to kill or inhibit the growth of bacteria, thereby helping the body fight off the infection. Antibiotics are the most commonly used antimicrobial agents, and they work by targeting specific structures or processes within bacteria to disrupt their ability to survive and reproduce.

When choosing an antimicrobial agent for a bacterial infection, it is essential to consider factors such as the type of bacteria causing the infection, the site of infection, the patient's medical history, and any allergies the patient may have. This information helps healthcare providers select the most effective antibiotic while minimizing the risk of adverse reactions or the development of antibiotic resistance.

Antibiotic resistance is a significant concern in the field of clinical bacteriology. It occurs when bacteria evolve mechanisms to resist the effects of antibiotics, making them less effective or completely ineffective in treating infections. To combat antibiotic resistance, healthcare providers must use antimicrobial agents judiciously, follow established guidelines for prescribing antibiotics, and educate patients about the importance of completing the full course of treatment.

In addition to antibiotics, other antimicrobial agents such as antivirals, antifungals, and antiparasitic drugs are used to treat infections caused by non-bacterial pathogens. Antivirals are used to treat viral infections such as influenza, HIV, and herpes, while antifungals are used to treat fungal infections such as candidiasis and aspergillosis. Antiparasitic drugs are used to treat infections caused by parasites such as malaria, giardiasis, and toxoplasmosis.

Overall, antimicrobial therapy is a vital component of the treatment of infectious diseases caused by bacteria. By understanding the principles of antimicrobial therapy, healthcare providers can effectively diagnose and treat bacterial infections while minimizing the risk of antibiotic resistance and promoting positive patient outcomes.

Prophylactic Measures

Prophylactic measures play a crucial role in preventing the spread of infectious diseases caused by bacteria. These measures are essential in reducing the risk of infection and maintaining public health. In this subchapter, we will discuss various prophylactic measures that can be implemented to prevent the transmission of bacterial infections.

One of the most effective prophylactic measures is vaccination. Vaccines are designed to stimulate the body's immune system to produce antibodies against specific bacteria, providing immunity against infection. Routine vaccinations, such as those for tetanus, diphtheria, and pertussis, are recommended to prevent bacterial infections in both children and adults. It is important for healthcare providers to educate patients about the importance of vaccination and encourage them to follow the recommended vaccination schedule.

Another important prophylactic measure is proper hygiene practices. Simple measures such as handwashing with soap and water, covering the mouth and nose when coughing or sneezing, and avoiding close contact with individuals who are sick can help prevent the spread of bacterial infections. Healthcare facilities should also implement strict infection control protocols to minimize the risk of healthcare-associated infections.

In certain situations, prophylactic antibiotics may be prescribed to individuals at high risk of developing bacterial infections. For example, individuals undergoing surgery or those with compromised immune systems may be given antibiotics to prevent postoperative infections or opportunistic infections. It is important for healthcare providers to use antibiotics judiciously and only when necessary to reduce the risk of antibiotic resistance.

Additionally, environmental control measures can help prevent the transmission of bacterial infections. This includes maintaining clean and sanitary living conditions, proper disposal of waste, and ensuring safe food handling practices. By addressing environmental factors that contribute to the spread of bacteria, we can reduce the risk of infection and protect public health.

In conclusion, prophylactic measures are essential in preventing the transmission of bacterial infections. By implementing vaccination, promoting proper hygiene practices, using antibiotics judiciously, and addressing environmental factors, we can reduce the risk of infection and protect individuals from bacterial diseases. It is important for healthcare providers and individuals to work together to implement these measures and promote public health.

Chapter 8

Prevention and Control of Bacterial Infections

Vaccines

Vaccines play a crucial role in the prevention and control of infectious diseases caused by bacteria. By stimulating the immune system to produce antibodies against specific bacteria, vaccines help protect individuals from becoming infected with these pathogens. Vaccines are composed of either weakened or killed bacteria, or components of bacteria, that are unable to cause disease but can still stimulate an immune response. This immune response results in the production of memory cells that can recognize and destroy the bacteria if the individual is exposed to them in the future.

One of the most well-known examples of a bacterial vaccine is the vaccine for tetanus, which is caused by the bacterium Clostridium tetani. The tetanus vaccine contains a toxin produced by the bacterium that is inactivated so it cannot cause disease. When a person is vaccinated against tetanus, their immune system recognizes the toxin as foreign and produces antibodies against it. These antibodies provide protection against tetanus infection by neutralizing the toxin if the person is exposed to it.

In addition to preventing infections in individuals, vaccines also play a critical role in controlling the spread of infectious diseases within communities. When a large proportion of a population is vaccinated against a particular bacterial pathogen, it creates a level of immunity known as herd immunity. This means that even individuals who are not vaccinated are protected because there are fewer opportunities for the bacteria to spread from person to person.

Despite their importance in preventing infectious diseases, vaccines are not without their challenges. Some bacterial pathogens are more difficult to develop vaccines against than others, either because they have complex structures that are hard to target with a vaccine, or because they can mutate rapidly and evade the immune response. Additionally, some individuals may have adverse reactions to vaccines, although these are typically mild and rare.

Overall, vaccines are a powerful tool in the fight against infectious diseases caused by bacteria. By stimulating the immune system to produce antibodies against specific bacterial pathogens, vaccines help protect individuals and communities from becoming infected. Continued research and development of new vaccines are essential to controlling the spread of bacterial infections and improving public health worldwide.

Infection Control Measures

Infection control measures are essential in preventing the spread of infectious diseases and maintaining a safe healthcare environment. In clinical bacteriology, it is crucial to implement strict protocols to minimize the risk of transmission of bacteria and other pathogens. These measures are designed to protect both patients and healthcare workers from acquiring infections and to prevent outbreaks within healthcare facilities.

One of the key infection control measures is proper hand hygiene. Healthcare workers must wash their hands regularly with soap and water or use alcohol-based hand sanitizers to prevent the spread of bacteria and viruses. Hand hygiene is particularly important before and after patient contact, before and after wearing gloves, and after touching potentially contaminated surfaces. By practicing good hand hygiene, healthcare workers can reduce the risk of transmitting pathogens to patients and prevent infections.

Another important infection control measure is the appropriate use of personal protective equipment (PPE). Healthcare workers should wear gloves, gowns, masks, and eye protection when caring for patients with infectious diseases to prevent exposure to pathogens. PPE should be used correctly and disposed of properly to avoid contamination. By following PPE guidelines, healthcare workers can protect themselves and their patients from acquiring infections.

Infection control measures also include environmental cleaning and disinfection. Healthcare facilities should have protocols in place for cleaning and disinfecting patient rooms, equipment, and common areas to prevent the spread of bacteria and viruses. Surfaces should be cleaned regularly using appropriate disinfectants to kill pathogens and reduce the risk of transmission. By maintaining a clean and sanitary environment, healthcare facilities can minimize the spread of infections and protect patients and staff.

In conclusion, infection control measures are essential in clinical bacteriology to prevent the spread of infectious diseases and maintain a safe healthcare environment. By implementing strict protocols for hand hygiene, personal protective equipment use, and environmental cleaning, healthcare facilities can reduce the risk of transmitting pathogens and prevent outbreaks. It is important for healthcare workers to follow these measures diligently to protect themselves, their patients, and the community from acquiring infections.

Public Health Strategies

Public health strategies are essential in the fight against infectious diseases caused by bacteria. These strategies focus on preventing the spread of bacteria, reducing the burden of disease, and promoting overall public health. One of the key public health strategies is vaccination, which helps protect individuals from bacterial infections by stimulating the immune system to produce antibodies against specific bacteria. Vaccination not only protects individuals but also helps to create herd immunity, reducing the overall prevalence of bacterial infections in the community.

Another important public health strategy is the implementation of infection control measures in healthcare settings. This includes practices such as hand hygiene, proper disinfection of medical equipment, and isolation of patients with infectious diseases to prevent the spread of bacteria from person to person. Infection control measures are crucial in preventing healthcare-associated infections, which can be caused by bacteria such as Staphylococcus aureus and Clostridium difficile.

Surveillance and monitoring of bacterial infections are also key components of public health strategies. By tracking the incidence and prevalence of bacterial infections, public health officials can identify trends, outbreaks, and emerging antibiotic resistance patterns. This information is essential for guiding public health interventions, such as targeted vaccination campaigns, antimicrobial stewardship programs, and public health education initiatives to raise awareness about the risks of bacterial infections and the importance of preventive measures.

Public health strategies also involve collaboration between public health agencies, healthcare providers, and the community. By working together, these stakeholders can develop and implement effective interventions to prevent and control bacterial infections. This includes sharing information, resources, and expertise to address the complex challenges of infectious diseases caused by bacteria. Public health strategies also involve engaging with policymakers to advocate for policies that support public health efforts, such as funding for research, surveillance, and prevention programs.

In conclusion, public health strategies play a critical role in the diagnosis and treatment of infectious diseases caused by bacteria. By focusing on prevention, surveillance, infection control, and collaboration, public health officials can effectively reduce the burden of bacterial infections and improve the overall health of the population. It is essential for healthcare professionals, policymakers, and the community to work together to implement and support these strategies to protect individuals and communities from the threat of bacterial infections.

Chapter 9

Emerging Trends in Clinical Bacteriology

Antibiotic Stewardship

Antibiotic stewardship is a crucial aspect of clinical bacteriology that aims to optimize the use of antibiotics in order to minimize the development of antibiotic resistance. This subchapter delves into the principles and strategies of antibiotic stewardship, highlighting the importance of prudent antibiotic use in combating infectious diseases.

One key principle of antibiotic stewardship is the concept of using the right antibiotic, at the right dose, and for the right duration. This involves selecting the most appropriate antibiotic based on the type of infection, the causative organism, and the patient's individual characteristics. By ensuring that antibiotics are prescribed judiciously, healthcare providers can help prevent the development of antibiotic resistance.

Another important aspect of antibiotic stewardship is the need to minimize unnecessary antibiotic use. This includes avoiding the overuse of antibiotics for conditions that are not bacterial in nature, such as viral infections. It also involves discontinuing antibiotics when they are no longer needed, in order to reduce the risk of adverse effects and the development of antibiotic resistance.

In addition to prescribing antibiotics appropriately, healthcare providers play a crucial role in educating patients about the importance of antibiotic stewardship. By explaining the rationale behind antibiotic prescriptions and discussing the potential risks of antibiotic resistance, patients can be empowered to take an active role in their own healthcare and make informed decisions about antibiotic use.

Overall, antibiotic stewardship is a multifaceted approach that involves healthcare providers, patients, and healthcare systems working together to ensure the responsible use of antibiotics. By following the principles of antibiotic stewardship, we can help preserve the effectiveness of antibiotics for future generations and combat the growing threat of antibiotic resistance.

Genomic Epidemiology

Genomic epidemiology is a rapidly evolving field within bacteriology that utilizes advanced sequencing technologies to study the genetic makeup of bacteria and other pathogens. By analyzing the entire genetic material of an organism, researchers can gain valuable insights into the transmission, evolution, and spread of infectious diseases. This powerful tool has revolutionized our understanding of how bacteria cause infections and how they can be controlled.

One of the key benefits of genomic epidemiology is its ability to track the transmission of bacteria in real-time. By comparing the genetic sequences of bacteria isolated from different patients, researchers can determine whether the infections are linked and identify potential sources of transmission. This information is crucial for implementing targeted interventions to prevent further spread of the disease and reduce the burden on healthcare systems.

In addition to tracking transmission, genomic epidemiology also helps researchers understand the evolution of bacteria and how they develop resistance to antibiotics. By studying the changes in the genetic code of bacteria over time, scientists can identify the mechanisms by which they acquire resistance genes and develop strategies to combat this growing threat. This information is essential for guiding antibiotic stewardship programs and developing new treatment options for drug-resistant infections.

Furthermore, genomic epidemiology plays a vital role in outbreak investigations and public health surveillance. By sequencing the genomes of bacteria isolated from outbreaks, researchers can identify the source of contamination, trace the spread of the infection, and implement control measures to prevent further cases. This proactive approach to outbreak management has been instrumental in containing the spread of infectious diseases and protecting public health.

Overall, genomic epidemiology represents a powerful tool in the fight against infectious diseases. By harnessing the power of advanced sequencing technologies, researchers can gain a deeper understanding of how bacteria cause infections, evolve, and spread. This knowledge is essential for developing effective strategies to control the spread of infectious diseases and protect the health of populations worldwide.

Novel Therapeutic Approaches

In recent years, the field of clinical bacteriology has seen significant advancements in the development of novel therapeutic approaches for the diagnosis and treatment of infectious diseases. These new approaches offer promising solutions to combat the growing threat of antibiotic resistance and emerging infectious diseases. In this subchapter, we will explore some of the latest innovations in the field of clinical bacteriology that are revolutionizing the way we diagnose and treat bacterial infections.

One of the most exciting developments in novel therapeutic approaches is the use of phage therapy. Phages are viruses that infect and kill bacteria, offering a targeted and highly specific approach to treating bacterial infections. Phage therapy has shown promising results in clinical trials, particularly for multidrug-resistant bacterial infections. By harnessing the power of phages, researchers are able to combat bacterial infections in a way that minimizes the risk of antibiotic resistance.

Another novel therapeutic approach that is gaining traction in clinical bacteriology is the use of probiotics. Probiotics are live bacteria and yeasts that are beneficial to human health, particularly in maintaining a healthy balance of gut microbiota. Recent studies have shown that certain strains of probiotics can help prevent and treat bacterial infections by promoting the growth of beneficial bacteria in the gut and inhibiting the growth of harmful bacteria. This approach offers a safe and effective alternative to traditional antibiotics for certain types of bacterial infections.

Advances in nanotechnology have also opened up new possibilities for the treatment of bacterial infections. Nanoparticles can be engineered to target and kill specific bacteria, offering a highly targeted and effective approach to treating infections. Nanoparticle-based therapies have shown promise in preclinical studies, demonstrating the potential to revolutionize the way we treat bacterial infections in the future.

In conclusion, the field of clinical bacteriology is rapidly evolving, with new therapeutic approaches offering exciting possibilities for the diagnosis and treatment of infectious diseases. From phage therapy to probiotics to nanotechnology, these novel approaches are providing innovative solutions to combat antibiotic resistance and address the challenges posed by emerging infectious diseases. As researchers continue to explore and develop these new therapies, the future of clinical bacteriology looks brighter than ever.

Chapter 10

Case Studies in Clinical Bacteriology

Case Study 1: Pneumonia

Pneumonia is a common and potentially serious respiratory infection that can be caused by a variety of bacteria, viruses, and fungi. In this case study, we will focus on a bacterial cause of pneumonia, specifically Streptococcus pneumoniae. This bacterium is a leading cause of community-acquired pneumonia and can also cause other serious infections such as meningitis and bacteremia.

The diagnosis of pneumonia caused by Streptococcus pneumoniae is often made based on clinical symptoms, such as fever, cough, and shortness of breath, as well as radiographic findings such as infiltrates on chest X-ray. Laboratory tests, such as sputum culture and blood culture, can help confirm the presence of the bacterium. In this case study, we will follow the diagnostic process for a patient presenting with symptoms consistent with pneumonia.

Treatment of pneumonia caused by Streptococcus pneumoniae typically involves antibiotics, with penicillin being the first-line choice. However, resistance to penicillin and other antibiotics is a growing concern, and alternative antibiotics may be necessary in some cases. We will discuss the appropriate antibiotic therapy for this patient, taking into account factors such as antibiotic resistance patterns and the severity of the infection.

In addition to antibiotic therapy, supportive care such as oxygen therapy and hydration are important components of treatment for pneumonia. In severe cases, hospitalization may be necessary to provide more intensive care. We will explore the management of this patient, including the need for hospitalization and monitoring for complications such as sepsis.

Overall, this case study will provide a comprehensive overview of the diagnosis and treatment of pneumonia caused by Streptococcus pneumoniae. By following the clinical course of this patient, readers will gain a better understanding of the challenges and considerations involved in managing this common infectious disease.

Case Study 2: Urinary Tract Infection

Urinary tract infections (UTIs) are a common and often painful condition that can affect individuals of all ages. In this case study, we will explore a patient presenting with symptoms of a UTI and the diagnostic process that led to the identification and treatment of the infectious agent responsible for the infection.

The patient, a 45-year-old female, presented to the clinic with complaints of burning sensation during urination, increased frequency of urination, and lower abdominal pain. Upon examination, the clinician noted cloudy urine and mild fever. Based on these symptoms, a presumptive diagnosis of UTI was made, and a urine sample was collected for laboratory analysis.

The urine sample was sent to the laboratory for a urinalysis and culture. The urinalysis revealed the presence of white blood cells and bacteria, confirming the presence of an infection in the urinary tract. A culture was performed to identify the specific bacterial species responsible for the infection. The culture results showed the growth of Escherichia coli, a common pathogen known to cause UTIs.

Treatment for the UTI was initiated with a course of antibiotics targeting the identified pathogen. The patient was advised to increase fluid intake and practice good hygiene to prevent future infections. Follow-up testing showed the resolution of symptoms and clearance of the infection, confirming the effectiveness of the chosen treatment.

This case study highlights the importance of accurate diagnosis and targeted treatment in the management of UTIs. By identifying the causative agent through laboratory testing, clinicians can tailor treatment to effectively combat the infection and prevent complications. Understanding the microbiology of UTIs is essential for healthcare professionals in providing optimal care for patients with these common infections.

Case Study 3: Sepsis

In this case study, we will be examining a patient with sepsis, a life-threatening condition that occurs when the body's response to an infection causes inflammation throughout the body. Sepsis can lead to septic shock, organ failure, and even death if not promptly diagnosed and treated. Understanding the pathogenesis of sepsis and identifying the causative organism are crucial steps in managing this condition.

The patient in this case study is a 65-year-old male with a history of diabetes and chronic obstructive pulmonary disease (COPD) who presents to the emergency department with a fever, low blood pressure, and altered mental status. These are all classic signs of sepsis, and prompt recognition of the condition is essential for initiating appropriate treatment. The patient's medical history and presenting symptoms suggest that he may have an underlying infection that has progressed to sepsis.

Upon admission, blood cultures are obtained to identify the causative organism. Blood cultures are an essential diagnostic tool in cases of suspected sepsis, as they can help determine the specific bacteria or fungi responsible for the infection. In this case, the blood cultures reveal the presence of a gram-negative bacterium, which is a common cause of sepsis in patients with COPD. The identification of the causative organism allows for targeted antibiotic therapy to be initiated promptly.

The patient is started on broad-spectrum antibiotics to cover the likely pathogens causing his sepsis. Antibiotic therapy is a critical component of sepsis management, as it helps to eradicate the infection and prevent further complications. In severe cases of sepsis, additional therapies such as intravenous fluids, vasopressors, and mechanical ventilation may be necessary to support the patient's vital functions. Close monitoring of the patient's clinical status and response to treatment is essential to ensure optimal outcomes.

In conclusion, sepsis is a serious condition that requires prompt recognition and aggressive treatment. By understanding the pathogenesis of sepsis, identifying the causative organism, and initiating appropriate therapy, healthcare providers can improve outcomes for patients with this life-threatening condition. This case study highlights the importance of a multidisciplinary approach to managing sepsis and underscores the critical role of bacteriology in the diagnosis and treatment of infectious diseases.

Chapter 11

Future Directions in Clinical Bacteriology

Technological Advancements

Technological advancements have revolutionized the field of clinical bacteriology, leading to significant improvements in the diagnosis and treatment of infectious diseases. One of the most impactful developments is the advent of molecular techniques such as polymerase chain reaction (PCR) and next-generation sequencing (NGS). These methods allow for the rapid and accurate identification of bacterial pathogens, enabling healthcare professionals to make informed decisions regarding patient care.

In addition to molecular techniques, advancements in imaging technology have also played a crucial role in the field of clinical bacteriology. Techniques such as computed tomography (CT) and magnetic resonance imaging (MRI) have greatly improved our ability to visualize and diagnose bacterial infections in various parts of the body. This non-invasive approach has led to earlier detection and more targeted treatment strategies, ultimately improving patient outcomes.

Another notable technological advancement in clinical bacteriology is the development of automated systems for microbial identification and antimicrobial susceptibility testing. These systems streamline the laboratory workflow, reducing the time and labor required for traditional culture-based methods. By providing rapid and accurate results, these systems help healthcare professionals select the most appropriate antibiotic therapy for patients, minimizing the risk of antibiotic resistance and improving treatment outcomes.

The integration of artificial intelligence (AI) and machine learning algorithms into clinical bacteriology has also shown great promise in recent years. These technologies can analyze vast amounts of data to identify patterns and predict bacterial behavior, aiding in the development of personalized treatment plans for patients. By harnessing the power of AI, healthcare professionals can optimize antibiotic therapy and improve patient care in a more efficient and cost-effective manner.

Overall, technological advancements in clinical bacteriology have significantly enhanced our ability to diagnose and treat infectious diseases. From molecular techniques to imaging technology to automated systems and AI, these innovations are transforming the field and improving patient outcomes. As technology continues to evolve, it is essential for healthcare professionals to stay informed and adapt their practices to incorporate these advancements into their daily clinical practice.

Global Health Implications

Global health implications of bacterial infections are significant and far-reaching. Infectious diseases caused by bacteria can spread rapidly across borders, impacting populations worldwide. The emergence of antibiotic-resistant bacteria poses a major challenge to healthcare systems around the world, as these pathogens are becoming increasingly difficult to treat. In addition, the lack of access to proper healthcare infrastructure and resources in many developing countries has led to higher rates of bacterial infections and poorer health outcomes.

One of the most pressing global health concerns related to bacterial infections is the rise of multidrug-resistant bacteria. These "superbugs" are able to evade the effects of multiple antibiotics, making treatment options limited and often ineffective. The spread of these resistant bacteria not only poses a threat to individual patients but also to public health as a whole. Efforts to combat antibiotic resistance through the development of new drugs and improved infection control measures are crucial to preventing the further spread of these dangerous pathogens.

Another global health implication of bacterial infections is the impact on vulnerable populations, such as children, the elderly, and those with compromised immune systems. These individuals are at higher risk of developing severe infections and experiencing complications from bacterial diseases. In many developing countries, lack of access to clean water, sanitation, and vaccines contributes to the higher burden of bacterial infections in these

A Primer for Healthcare Professionals

populations. Addressing these disparities and improving access to healthcare services is essential for reducing the global impact of bacterial infections.

Globalization has also played a role in the spread of bacterial infections, as increased travel and trade have made it easier for pathogens to move between countries. Outbreaks of infectious diseases such as tuberculosis, cholera, and meningitis can quickly become global health emergencies if not contained quickly and effectively. Collaboration between countries and international organizations is essential for monitoring and responding to these outbreaks, as well as for developing strategies to prevent the spread of bacterial infections on a global scale.

In conclusion, bacterial infections have significant implications for global health, affecting populations worldwide and posing challenges to healthcare systems and public health efforts. Addressing the rise of antibiotic-resistant bacteria, improving access to healthcare services for vulnerable populations, and strengthening international cooperation are essential steps in combating the global impact of bacterial infections. By working together to prevent the spread of infectious diseases and improve treatment options, we can help protect the health and well-being of people around the world.

Research Priorities

In the field of clinical bacteriology, research priorities play a crucial role in advancing our understanding of infectious diseases and improving diagnostic and treatment strategies. By focusing on key areas of investigation, researchers can identify new pathogens, develop novel diagnostic tests, and discover more effective treatments for bacterial infections.

One of the main research priorities in clinical bacteriology is the identification of emerging infectious diseases. With the rise of globalization and increased travel, new pathogens are constantly being introduced into different populations. By monitoring trends in bacterial infections and studying their genetic makeup, researchers can better prepare for potential outbreaks and develop targeted interventions to prevent the spread of these diseases.

Another important research priority is the development of rapid diagnostic tests for bacterial infections. Traditional methods of diagnosing bacterial infections can be time-consuming and labor-intensive, leading to delays in treatment and potential complications for patients. By investing in research to create faster and more accurate diagnostic tools, healthcare providers can quickly identify the causative agent of an infection and prescribe the most appropriate treatment.

Additionally, research in clinical bacteriology is focused on understanding the mechanisms of antibiotic resistance in bacteria. The overuse and misuse of antibiotics have led to the emergence of resistant strains, making it increasingly difficult to treat bacterial infections. By studying how bacteria develop resistance to antibiotics and identifying new targets for drug development, researchers can help combat this growing threat and ensure that effective antibiotics remain available for future generations.

Finally, research priorities in clinical bacteriology also include investigating the role of the microbiome in health and disease. The human microbiome, which consists of trillions of bacteria living in and on the body, plays a crucial role in maintaining our overall health and immune function. By studying how changes in the microbiome can impact susceptibility to bacterial infections, researchers can develop new strategies to promote a healthy microbial balance and prevent disease. By addressing these key research priorities, the field of clinical bacteriology can continue to make significant strides in diagnosing and treating infectious diseases, ultimately improving patient outcomes and public health.

Conclusion: The Future of Clinical Bacteriology

In conclusion, the future of clinical bacteriology is promising as advancements in technology continue to revolutionize the field. With the development of rapid diagnostic tools such as molecular techniques and next-generation sequencing, healthcare professionals can now identify pathogens with greater speed and accuracy than ever before. This allows for targeted treatment strategies and the potential to reduce the spread of infectious diseases.

Furthermore, the increasing prevalence of antibiotic resistance poses a significant challenge for clinical bacteriology. As more bacteria become resistant to traditional antibiotics, it is crucial for researchers to develop new antimicrobial agents to combat these evolving pathogens. Additionally, the implementation of antimicrobial stewardship programs can help to promote the appropriate use of antibiotics and reduce the development of resistance.

Another important aspect of the future of clinical bacteriology is the continued emphasis on infection prevention and control measures. By implementing strict protocols for hand hygiene, environmental cleaning, and the appropriate use of personal protective equipment, healthcare facilities can prevent the spread of nosocomial infections and protect both patients and healthcare workers.

Moreover, the integration of artificial intelligence and machine learning algorithms into clinical bacteriology has the potential to revolutionize the way infectious diseases are diagnosed and treated. These technologies can analyze vast amounts of data to identify patterns and trends that may not be apparent to human clinicians, leading to more accurate and personalized treatment plans for patients.

In conclusion, the future of clinical bacteriology holds great promise for improving the diagnosis and treatment of infectious diseases. By embracing new technologies, combating antibiotic resistance, and prioritizing infection prevention and control measures, healthcare professionals can work towards a future where infectious diseases are effectively managed and controlled.

References

In the realm of clinical bacteriology, references play a crucial role in providing evidence-based information to support diagnosis and treatment decisions. This subchapter on references aims to guide healthcare professionals and researchers in accessing the most up-to-date and relevant literature in the field of infectious diseases.

The references section of this book includes a comprehensive list of primary sources, such as peer-reviewed journal articles, textbooks, and clinical guidelines. These sources have been carefully selected to ensure the accuracy and reliability of the information presented in this book. By consulting these references, readers can delve deeper into specific topics and gain a more thorough understanding of the latest advancements in clinical bacteriology.

For healthcare professionals specializing in bacteriology, having access to a robust collection of references is essential for staying current with the rapidly evolving landscape of infectious diseases. By referencing reputable sources, clinicians can make informed decisions when diagnosing and treating patients with bacterial infections. Additionally, researchers can use references to support their own studies and contribute valuable insights to the field of bacteriology.

The references provided in this subchapter cover a wide range of topics related to clinical bacteriology, including microbial pathogenesis, antimicrobial resistance, and diagnostic methods. These sources have been carefully curated to encompass both foundational knowledge and cutting-edge research in the field. By consulting these references, healthcare professionals can enhance their clinical practice and contribute to the advancement of bacteriology as a whole.

In conclusion, the references section of this book serves as a valuable resource for healthcare professionals and researchers seeking to expand their knowledge and expertise in clinical bacteriology. By utilizing these references, readers can access high-quality information that is essential for making informed decisions in the diagnosis and treatment of infectious diseases. Ultimately, the references provided in this subchapter are a testament to the importance of evidence-based practice in the field of bacteriology.

Dear Reader,

Thank you for choosing to read "Clinical Bacteriology: Diagnosis and Treatment of Infectious Diseases." I hope this book has provided you with valuable insights and practical knowledge to enhance your understanding and management of infectious diseases.

Your feedback is incredibly important to me and to other healthcare professionals seeking reliable resources. If you found this book helpful or informative, I would greatly appreciate it if you could take a moment to leave a review on Amazon. Your honest review will help others discover the book and understand its value.

To leave a review, simply follow these steps:

1. Go to the book's page on Amazon.

2. Scroll down to the "Customer Reviews" section.

3. Click on "Write a customer review."

4. Share your thoughts and experiences with the book.

Thank you so much for your time and support. Your review means a lot to me and to the healthcare community!

Best regards,

Bhupen Thapa

www.ingramcontent.com/pod-product-compliance
Lightning Source LLC
Chambersburg PA
CBHW072019230526
45479CB00008B/298